The Polar Regions

Jennifer Prior, Ph.D.

Consultants

Andrea Johnson, Ph.D.
Assistant Professor of History
California State University, Dominguez Hills

Eileen Marthiensen, M.Ed.
Teacher, Alberta, Canada

Brian Allman
Principal
Upshur County Schools, West Virginia

Publishing Credits

Rachelle Cracchiolo, M.S.Ed., *Publisher*
Emily R. Smith, M.A.Ed., *SVP of Content Development*
Véronique Bos, *Vice President of Creative*
Dani Neiley, *Editor*
Fabiola Sepulveda, *Series Graphic Designer*

Image Credits: p.5 Shutterstock/I. Noyan Yilmaz; p.9 (bottom) Flickr/MerryJack/AustralianMuseum; p.10 Shutterstock/Robert Szymanski; p.11 Louis Choris; p.13 (bottom) Shutterstock/Georg Kristiansen; p.14 Alamy/Robertharding; p.15 (top) Gaelen Marsden; p.15 (bottom) Getty Images/Andrew H. Walker; p.21 (top) Alamy/Sue Clark; p.25 (top) Getty Images/AGF; p.27 (top) Alamy/Design Pics Inc; p.27 (bottom) NASA/Joshua Stevens; all other images from iStock and/or Shutterstock

Library of Congress Cataloging-in-Publication Data

Names: Prior, Jennifer Overend, 1963- author.
Title: The polar regions / Jennifer Prior.
Description: Huntington Beach : Teacher Created Materials, Inc, 2023. | Includes index. | Audience: Ages 8-18 | Summary: "The extreme cold of the polar regions may seem uninviting. But people and animals prosper in these frigid environments. With seasons of unending daylight and enduring darkness, the region is called home by people whose ancestors have thrived for centuries. These are lands of timeless traditions and snow-covered beauty"-- Provided by publisher.
Identifiers: LCCN 2022038397 (print) | LCCN 2022038398 (ebook) | ISBN 9781087695211 (paperback) | ISBN 9781087695372 (ebook)
Subjects: LCSH: Polar regions--History--Juvenile literature. | Indigenous peoples--Arctic Regions--Juvenile literature.
Classification: LCC G580 .P75 2023 (print) | LCC G580 (ebook) | DDC 998--dc23/eng20221107
LC record available at https://lccn.loc.gov/2022038397
LC ebook record available at https://lccn.loc.gov/2022038398

Shown on the cover is Lofoten, Norway.

This book may not be reproduced or distributed in any way without prior written consent from the publisher.

5482 Argosy Avenue
Huntington Beach, CA 92649
www.tcmpub.com
ISBN 978-1-0876-9521-1
© 2023 Teacher Created Materials, Inc.

Table of Contents

Land of Extremes 4
Arctic and Antarctic Features 6
Indigenous Peoples of the Arctic 10
People in Antarctica 14
Polar Wildlife 18
Surviving Extreme Temperatures 22
There's No Place Like Home 26
Map It! 28
Glossary 30
Index 31
Learn More! 32

iceberg in Wilhemina Bay, Antarctica

Land of Extremes

Of all the regions in the world, the polar regions may be among the most extreme due to their cold temperatures. The polar regions are the areas around the North Pole and the South Pole. These are referred to as the Arctic Circle and the Antarctic Circle. From the ways of life to the animals, the Arctic and Antarctic regions are truly unique.

Because Earth is tilted on an **axis**, night and day are affected in the polar regions. During the winter months, the sun is almost never seen. During the summer, the sun almost never sets, so life is lived in unending daylight. Can you imagine the sun being out at midnight? People who are not used to this have to cover their windows so they can sleep.

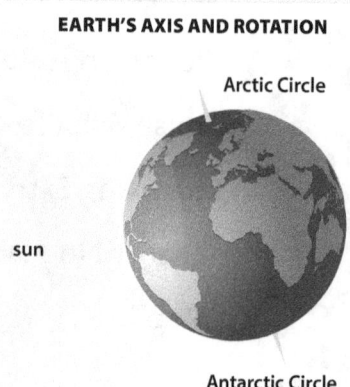

EARTH'S AXIS AND ROTATION

Arctic Circle

sun

Antarctic Circle

In Tromsø, Norway, between late November and mid-January, it is dark for more than 20 hours a day.

scientists in Antarctica

Polar extremes require **keen** skills for long-term survival. People have developed important skills for how they dress, hunt, and create shelters. Without these skills, they would not be able to survive. Animals have unique characteristics that help them withstand the cold. And while these **remote** polar regions can be dangerous, they also have beautiful landscapes and unique features.

The Great Ice Age

There have been several ice ages throughout history. The most recent global ice age ended roughly 11,700 years ago. Long, long ago, ice covered one-third of Earth. This happened because the planet was much colder. Earth warmed up over time. Ice melted and left many of the glaciers we see today in the Arctic.

Arctic and Antarctic Features

The Arctic and Antarctic regions share some characteristics. Cold weather is one thing they share! But each region is unique. Let's learn about these polar places.

On Top of the World

The Arctic consists of large land masses and thick floating ice. The northernmost parts of several countries are in the Arctic Circle. These are Canada, Greenland, Norway, Sweden, Finland, and Russia. The state of Alaska (part of the United States) is also partially in the Arctic Circle.

Certain places with higher **elevations** in the Arctic have permanent snow and ice. Some grasses and shrubs grow. But there are no trees. This **terrain** is known as the tundra.

You might be wondering if it ever warms up in the Arctic. Temperatures in the summer reach a high of about 38–45 °F (3–7 °C). That's pretty chilly. In winter, the average high is –30 °F (–34 °C). The coldest temperature ever recorded in the Arctic Circle was –93.3 °F (–69.6 °C)! That was in Greenland in 1991.

Because of these extremely cold temperatures, much of the Arctic Circle has permafrost. This is ground that is always frozen. The surface is not necessarily covered in snow and ice, but the ground below stays frozen all year long. With the ground permanently frozen, water does not drain well in this region. For that reason, water builds up and forms shallow lakes that can be seen all over the Arctic.

frosted tundra of the Kola Peninsula, Russia

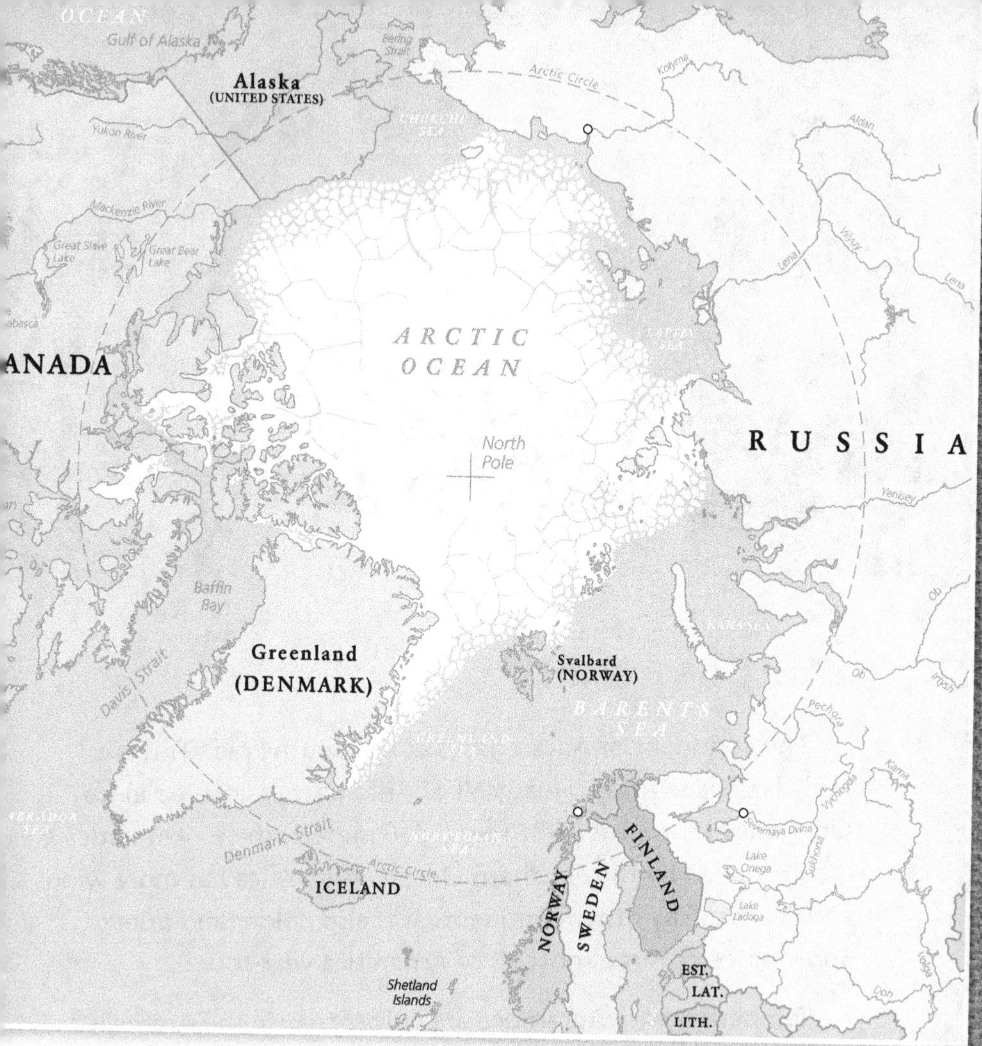

The Bear

The word *arctic* comes from the Greek word *arktos*, which means "bear." It was given that name from two **constellations** of stars called the *Great Bear* and *Little Bear*. They are also called the *Big Dipper* and *Little Dipper*. If you can find the two stars at the end of the Big Dipper, they will point you toward the tail of the Little Dipper. The star at the end of the Little Dipper's tail is known as the North Star. In the past, sailors and Inuit used the North Star to navigate oceans and seas at night.

Paradise Bay, Antarctica

Way Down South

The continent of Antarctica is at the South Pole. This is a fairly large land mass. Nearly all of Antarctica is covered in ice. On average, the ice is 7,087 feet (2,160 meters) thick. Antarctica is surrounded by the Southern Ocean. Antarctica has more wind and ice than any other continent. It is also colder and drier. Snow or ice covers almost all of Antarctica year-round.

Temperatures in Antarctica are colder than in the Arctic in the winter. In fact, in 1983, the low reached −128.6 °F (−89.2 °C)! In the summer, most parts of the region still have freezing temperatures. But on the Antarctic Peninsula, it can get as warm as 59 °F (15 °C).

If cold temperatures were not extreme enough, factor in the **wind chill**. When wind blows, it makes the air feel colder than it is. That's because skin loses heat when wind hits it. Winds in this region can reach up to 200 miles (320 kilometers) per hour.

Surprisingly enough, Antarctica is considered a desert! This is because it gets very little moisture. It only gets between 2 and 10 inches (5 and 25 centimeters) of snow per year. And it almost never rains.

Fossil Finds

Fossils found in Antarctica tell us that it was not always as cold as it is now. Long, long ago, there were reptiles and amphibians. There were even dinosaurs. Scientists have also found evidence of ferns and trees.

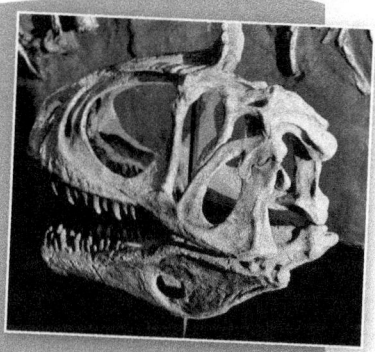

Cryolophosaurus skull

Indigenous Peoples of the Arctic

There are many **Indigenous** groups who live in the Arctic. Each group has a unique connection to the land. Each group has its own language and ways of obtaining the things they need for survival.

Inuit

In the past, Inuit depended on the land for survival. When they hunted seals and whales, they were careful to use every part of the animal so that nothing went to waste.

In the summer, Inuit made homes from animal skins that they stretched over wood or whale bones. In winter, they made traditional igloos out of blocks of snow. These homes stayed warm throughout the season. This is because heat rises, and cold air sinks. Body heat warmed the air and kept all but the bottom of the igloo warm.

Today, many Inuit live in homes in cities and towns in Alaska, Canada, and Greenland. Some work in mines and oil fields. It is estimated that there are 180,000 Inuit in the world.

an Inuit girl with her two puppies

An Unangan man paddles past some seals.

Unangax̂

The Unangax̂ have lived in western Alaska and the Aleutian Islands for thousands of years. They are also known as the Aleuts. The Unangax̂ were skilled for survival in this area. They hunted animals in the water and on land, including seals, whales, caribou, and bears. They used wild grasses to weave baskets and used stone, bone, or ivory to make containers, needles, and other objects. Today, it is estimated that 15,000 people have Unangan ancestors.

Speaking Out

Alexandria Abuzanuq Ivanoff is a young Inuit woman from Alaska. She works as a journalist. She speaks out for the rights of people in the Arctic. She also uses her writing to share about the importance of caring for the Arctic ecosystem.

baby reindeer in a Khanty camp in Russia

Khanty

The Khanty people live in the northern part of Russia called *Siberia*. In the area where they live, winter lasts for six months. At times, there can be 6.5 feet (2 meters) of snow on the ground. They use sleds and skis to get around.

The Khanty herd reindeer. They have done this since the 15th century. The Khanty follow the reindeer for part of the year. When they are traveling, they live in tents called *chums*. They occasionally stay in one location to raise the reindeer. They use reindeer meat for food, but the rest of the animal does not go to waste. The people use the hide for clothing, tent covers, and shoes. They use the bones and other parts of the animal for making tools.

Sami

The Sami people live in the northern parts of Norway, Sweden, Finland, and Russia. The Sami descend from nomadic peoples in northern Scandinavia. In the past few decades, the countries where Sami live have created **policies** to help protect the people and their land.

Like the Khanty, many of the Sami are reindeer herders. Reindeer herding is no longer the basis of their economy, though. A lot of Sami are fishers, farmers, and miners, while others choose to work and live in cities and towns.

A Sami man works on his sleigh in Lapland, Finland.

Arctic People

You might think few people live in the Arctic, but about four million people actually do. Only about 10 percent of them are Indigenous. A lot of people move to the Arctic to live and work.

People in Antarctica

There are no permanent residents in Antarctica. However, hundreds of people can be there at any given time. Scientists from all over the world live there for periods of time. They conduct research and report their findings. And tourists visit to see the amazing sights and wildlife.

In 1959, 12 countries initially agreed that the land would be used for research. This is called the *Antarctic Treaty*. There are now 70 research stations from 29 countries on this island continent. The researchers have all agreed to work together and take care of the natural environment.

Scientists there are interested in studying many things in Antarctica, including the animals. They study snow, ice, and rocks. Some study the skies. Others study the ocean. And some study the weather.

Port Lockroy in Antarctica was once a research station and is now open to tourists and visitors during the summer.

McMurdo Station is the largest Antarctic research station.

Some of the stations are very small and can only fit a few people. Other stations, such as McMurdo Station of the United States, are large. McMurdo Station has around 100 buildings, several airplane landing strips, and a helicopter pad.

Not all people live in Antarctica for science. Some come for the Antarctic Artists and Writers Program. Writers, painters, photographers, and other artists can apply to the program. Its purpose is to increase understanding of the region and the work that is done there.

Ice People

Ann Aghion is a French American filmmaker. She was selected to be part of the Antarctic Artists and Writers Program. She spent four months exploring the region and interviewing researchers. She made a film called *Ice People* about her experience.

Remote Living

In the past, people used to hunt and gather everything they needed for survival. Trade and **imports** have allowed people to get things they need more easily. But with the Arctic being so far north, there are some unique challenges.

Some towns in the Arctic cannot be **accessed** by roads. Utqiagvik, Alaska, is one of them. It is the northernmost town in the United States. You can only get there by plane or by boat. This is because it is in such a remote location. Other places in the Arctic are only accessible by plane. That is how you can get to Nord Station, Greenland. There is too much sea ice to make the journey by boat.

Northern Lights

The northern lights are an amazing sight to see. They are also called *aurora borealis*. They can be seen at the North Pole and across the Arctic Circle. They can be seen in certain places in Canada and sometimes even in the northern part of the United States. These dancing lights are caused when electrically charged **particles** from the sun enter Earth's atmosphere. They are best seen in the winter when the sky is dark.

Utqiaġvik, Alaska

Being more difficult to access means that basic supplies are harder to get. Supplies and other goods often cost a lot of money because of how hard it is to deliver them. At grocery stores in the Arctic Circle, fresh produce can be expensive. Can you imagine paying $37.00 for half a watermelon? This is how much one cost in Utqiaġvik in October 2016!

In Search of Adventure

Tourism is a key part of the economies in the polar regions. People come from all over the world. They visit for the chance to see glaciers, whales, and the northern lights.

Another thing that brings people to the Arctic is jobs. Several industries are active in this area. This includes fishing and mining. Some people work as tour guides.

cod fishing boat

Polar Wildlife

Despite the freezing temperatures, many animals call the polar regions home. Their bodies are designed to **adapt** to these climates.

Animals of the Arctic

Polar bears are the largest of all bears on the planet. They live only in the Arctic region. Polar bears have been around for a long time. The oldest known polar bear fossil is between 110,000 and 130,000 years old! Scientists were able to determine this based on the location of the fossil in the earth.

Polar bears spend most of their time in **frigid** ocean water or on sea ice. A polar bear's body has 4 inches (10 centimeters) of fat under the skin. The fat and its two thick layers of fur help keep it warm. Polar bears spend half their time hunting, and they mostly eat seals. Seals are high in fat, which polar bears need to survive.

polar bears

Arctic foxes live in the polar region as well. They feed on smaller animals, such as **lemmings** and birds. To blend in with their surroundings, Arctic foxes have grayish-brown coats in the summer. Their fur is white in the winter.

Arctic fox

Pacific walruses live in the Arctic, too. They have thick skin, and both males and females have long tusks. Walruses use their tusks to pull themselves up onto ice or rocks. They also use their tusks to fight off other animals.

Bowhead Whales

In the nineteenth century, sailors hunted bowhead whales for their oil, blubber, and **baleen**. In recent years, the species was protected under the Endangered Species Act. The number of bowhead whales has increased since then. The International Whaling Commission also exists to protect whales. Nearly 90 countries around the world participate in the international effort.

Animals of Antarctica

All animals in Antarctica are considered marine wildlife. That means they live in and around the ocean. There are no polar animals that live entirely on land in Antarctica.

The penguin is one of the most loved polar animals. These birds have flippers instead of wings. So they do not fly. Instead, they use their flippers to swim. They can swim up to 15 miles (24 kilometers) per hour when chasing after food. While there are many different kinds of penguins, the Emperor penguin is the largest. It is more than 4 feet (1.2 meters) tall.

There are six different kinds of seals that live in Antarctica. Seals move awkwardly on land, but they are great swimmers. They mostly eat fish, **krill**, and squid, but some seals eat penguins. The elephant seal is the largest. It can weigh more than 11,000 pounds (5,000 kilograms).

elephant seal and king penguins

An orca pod surrounds a seal.

Orcas are found in many oceans around the world, but a large number of them live in the Southern Ocean. They eat some fish, but they mainly eat seals. Seals have a lot of fat on their bodies, and the whales need fat to survive the freezing temperatures of Antarctic waters.

Watch Out!

Penguins have a lot of **predators** to watch out for. Seals, sharks, and orcas feast on them. Every time penguins go for a swim, they have to be careful, or they might become a meal!

Surviving Extreme Temperatures

How do people live in extremely cold environments? Without preparation and skills, it can be very dangerous—and even deadly.

How the Body Responds to Cold

The body's main goal in extreme cold is to survive. It sends heat to the most important parts. Organs such as the brain, heart, lungs, and kidneys are essential to life. Extremities, such as fingers, toes, arms, and legs, are not as important. So, when too much heat moves away from toes, for example, they get frostbite. It is common for a person to lose fingers, toes, feet, or even the end of their nose from frostbite.

Brrrr!

Shivering is one way the body tries to stay warm. When muscles move, they create heat. You have probably noticed that your body warms up when you run around outside, even if it's cold. Shivering is one thing your body does for survival.

Initially, when exposed to cold, skin turns red. Have you ever noticed that your cheeks and nose get red in the cold? That is caused by blood vessels **dilating**. If the temperature is too cold or if you stay outside too long, the blood vessels **constrict**, and the blood moves away from the skin's surface.

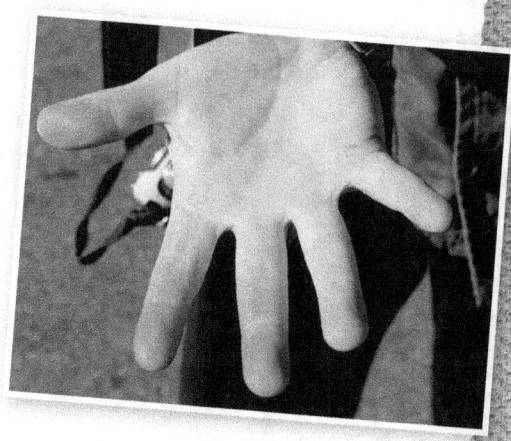

Hypothermia is a dangerous condition that happens when the core of the body does not have enough heat. It begins when the body is losing heat faster than it is producing it. This can damage organs, and it can even lead to death.

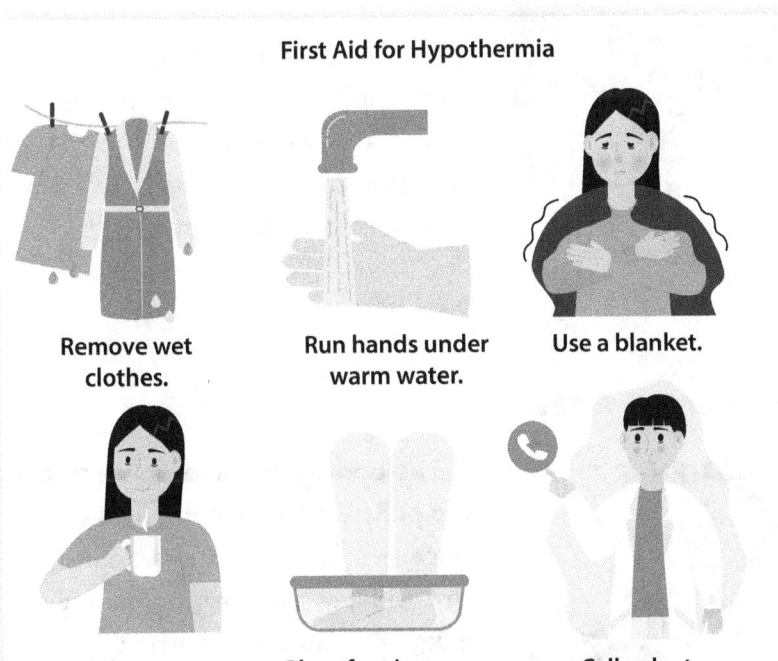

First Aid for Hypothermia

- Remove wet clothes.
- Run hands under warm water.
- Use a blanket.
- Drink a warm beverage.
- Place feet in warm water.
- Call a doctor.

Cold weather brings many dangers. So, how do people survive in the polar regions? Whether living in a house or a tent, or wearing modern clothing or animal skins, these strategies are important to keep in mind.

Stay Out of Wind

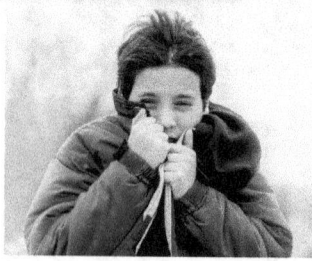

When exposed to wind, skin quickly loses heat. For this reason, it is vital to find protection from the wind to stay warm.

Stay Dry

One of the quickest ways to lose body heat is being wet. Avoid being in water or wearing wet clothes. Using a waterproof jacket is ideal. This can help prevent hypothermia.

Protect Fingers and Toes

The body's natural response to cold is to pull heat to the core. This can be dangerous for fingers, toes, and even arms and legs. So, wearing warm gloves, socks, and boots is important.

An Inuit boy stays warm in clothing made from animal skins and fur.

Stay Hydrated

Surprisingly enough, drinking water helps keep the body warm. The circulation of blood creates body heat, and water helps to increase the amount of blood in the body.

Dress in Layers

Survivalists recommend dressing in four layers. People should have long underwear, pants and a shirt, a coat, and a wind jacket. Layers keep the body warm and dry. They help to keep body heat from escaping.

Fabric Types

Fabric types are crucial in cold weather. Cotton should be avoided. Cotton layers can trap moisture, such as body sweat. This will end up making a person feel colder. Instead, using polyester or silk is better. These fabrics allow moisture to evaporate. Lastly, waterproof fabrics should be used. These help keep people dry in rain and snow.

Arctic summer in Svalbard, Norway

There's No Place Like Home

The snow-covered beauty of the polar regions is a sight to behold. Visitors come to see the uniqueness of the land. They come to see the untouched wilderness and the northern lights. Scientists endure the extreme cold to gather information about the land, animals, oceans, and sky. Others come to film, paint, and write about what they see.

And while the terrain can be fierce and even dangerous, it is a welcome home to the many people and animals who thrive in the environment. For thousands of years, Indigenous peoples in the Arctic have lived off the land. They used the land's natural resources for their shelter and clothing. They used survival skills passed down to them from their ancestors.

Traces of these traditions can be seen throughout the Arctic. Traditional clothing, storytelling, and artwork paints a picture of the past. But towns and cities in the polar regions are just as modern as the rest of the world. Resources are imported and exported. Many people in the Arctic work in industries that benefit countries around the world.

Arctic hare

The polar regions are lands of extremes, especially when it comes to climate. But to the people and animals who live there, the polar regions are home.

A Yupik artist tells a traditional story to children.

Melting Ice

Scientists have learned that ice in the Arctic is melting at a rate of 13 percent per decade. This is caused by global warming. Disappearing ice can have harmful effects on the animals that live in the Arctic. Polar bears need the ice to hunt other animals. Walruses need the ice to rest and find their food. Also, the melting ice can harm the rest of the planet, too. Melting ice raises sea levels. This can affect coastal cities and towns.

Map It!

Many countries have land that falls in the Arctic Circle. With a group, create a map of a country that shows where its polar region and other landmarks are located.

1. Choose one of the countries in the Arctic Circle: United States, Canada, Greenland, Iceland, Norway, Finland, Sweden, or Russia.
2. Make a map of it on poster board. Make sure to highlight, color in, or shade its polar region.
3. Do research online to identify and label towns, landmarks, and surrounding oceans or landmasses. Add these to your map.
4. Find out whether any Indigenous peoples live on the land. Label their names and the areas where they live on your map.
5. Share your map with the class.

Saqqaq, Greenland

Glossary

accessed—entered

adapt—to change or make adjustments to behavior so that it is easier to function

axis—an imaginary line that something spins around

baleen—a tough material that hangs down from the upper jaw of whales and can filter small ocean animals out of seawater

constellations—groupings of stars that form patterns

constrict—to become more narrow

dilating—becoming wider

elevations—the heights of things above sea level

frigid—extremely cold

imports—goods and resources that are brought into a place or region from somewhere else

Indigenous—from or native to a particular area

keen—highly developed

krill—small, shrimp-like sea creatures

lemmings—small rodents

particles—small bits of matter

policies—rules or laws

predators—animals that hunt and eat other animals

remote—far, out-of-the-way, or secluded from other things

terrain—land

wind chill—temperature that shows how cold the air feels on skin when wind is factored in

Index

Alaska, 6–7, 10–11, 16–17, 27

Aghion, Ann, 15

Antarctica, 5, 8–9, 14–15, 20–21

Antarctic Artists and Writers Program, 15

Antarctic region, 4–5, 8–9, 14–15, 20–21, 27

Arctic region, 4–8, 10–11, 13, 16–19, 26–27

aurora borealis, 16

Canada, 6–7, 10, 16

Finland, 6–7, 13

Greenland, 6–7, 10, 16

Iceland, 7

Inuit, 7, 10–11, 25

Khanty, 12–13

McMurdo Station, 15

North Pole, 4, 16

northern lights, 16–17, 26

Norway, 4, 6–7, 13, 26–27

Russia, 6–7, 12–13

Sami, 13

South Pole, 4, 8–9

Sweden, 6–7, 13

Unangax̂, 11

Utqiaġvik, 16–17

reindeer in Sweden

Learn More!

Ann Bancroft is a world-famous polar explorer. She started the Ann Bancroft Foundation as a way of inspiring women and girls to achieve their dreams. Research her life. Use the information you find to create a news report about her. (Use the search term "Ann Bancroft explorer" so you don't confuse her with an actress by the same name.)

- Find two interesting facts about Bancroft when she was younger.
- What amazing achievement did Bancroft accomplish in February 2021?
- What are Bancroft's other accomplishments?
- Describe one of Bancroft's hobbies.
- What does the Ann Bancroft Foundation do?

humpback whales near Greenland

www.ingramcontent.com/pod-product-compliance
Lightning Source LLC
Chambersburg PA
CBHW050450010526
44118CB00013B/1772